My Favorite Animal

I LOVE SLOTHS!

By Beth Gottlieb

Gareth Stevens
PUBLISHING

Please visit our website, www.garethstevens.com. For a free color catalog of all our high-quality books, call toll free 1-800-542-2595 or fax 1-877-542-2596.

Library of Congress Cataloging-in-Publication Data
Names: Gottlieb, Beth, author.
Title: I love sloths! / Beth Gottlieb.
Description: Buffalo, New York : Gareth Stevens Publishing, [2023] | Series: My favorite animal | Includes index.
Identifiers: LCCN 2022014033 (print) | LCCN 2022014034 (ebook) | ISBN 9781538283431 (paperback) | ISBN 9781538283455 (library binding) | ISBN 9781538283462 (ebook)
Subjects: LCSH: Sloths–Juvenile literature.
Classification: LCC QL737.E2 G63 2023 (print) | LCC QL737.E2 (ebook) | DDC 599.3/13–dc23/eng/20220518
LC record available at https://lccn.loc.gov/2022014033
LC ebook record available at https://lccn.loc.gov/2022014034

First Edition

Published in 2023 by
Gareth Stevens Publishing
2544 Clinton Street
Buffalo, NY 14224

Editor: Kristen Nelson
Designer: Rachel Rising

Photo credits: Cover, Lukas Kovarik/Shutterstock.com; pp. 5, 9, 17, 24 Damsea/Shutterstock.com; p. 7 Kjersti Joergensen/Shutterstock.com; p. 7 rvsphoto/Shutterstock.com; pp. 11, 24 Penty Photography/Shutterstock.com; p. 13 kungverylucky/Shutterstock.com; p. 15 Inspired By Maps/Shutterstock.com; p. 19 J_K/Shutterstock.com; p. 21 worldswildlifewonders/Shutterstock.com; p. 23 Sofie Hojabri/Shutterstock.com.

Printed in the United States of America

CPSIA compliance information: Batch #CWGS23: For further information contact Gareth Stevens, New York, New York at 1-800-542-2595.

Contents

My favorite animal
is a sloth!

There are six kinds.

They all have fur.

They have long claws.

They live in rain forests.
They live up in the trees.

They hang
from branches.

They climb.
They move slowly.

They eat leaves.
They may eat fruit.

They sleep a lot!

I love sloths!

Words to Know

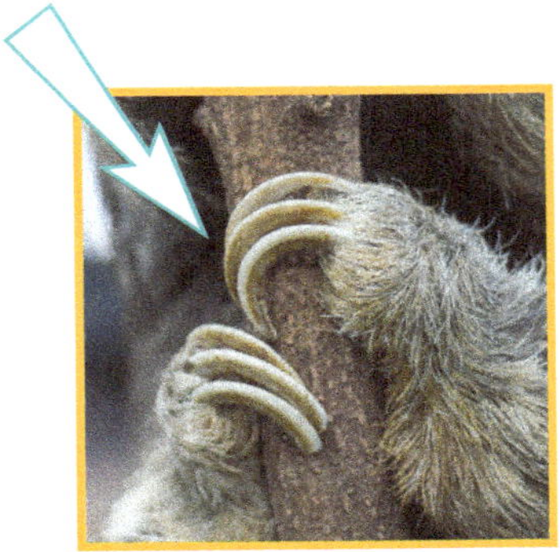

claws

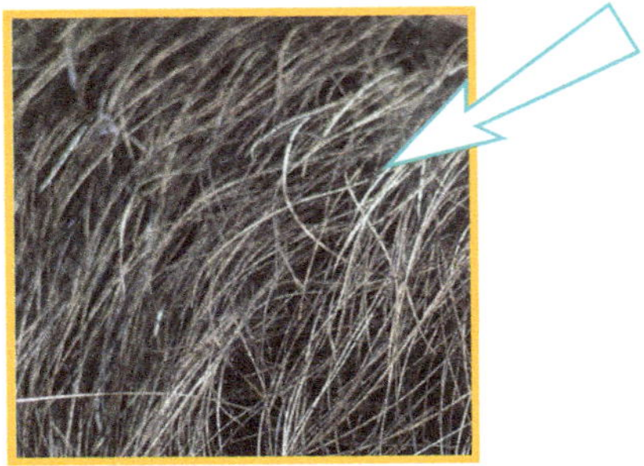

fur

Index